CW00531634

Higher Grade Biology
Revision Notes

Denyse Kozub

Published by   Leckie & Leckie
St. Andrews

Tel. 01334 - 475656
Fax. 01334 - 477392

Edited by   Morag Barnes, Dave Mole and
Andrew Morton

A CIP Catalogue record for this book is available
from the British Library.

ISBN 1-898890-35-8
2nd edition

Leckie & Leckie

Contents

1. Cell Biology

Cell Variety in Relation to Function

Not all cells are identical. The shape and contents of individual cells show differences related to function.

1. Some Specialised Animal Cells

nerve cells (neurones)	muscle cells	leucocytes (white blood cells)
the extensions make them suitable for carrying electrical impulses	these cells are elongated and can bring about movement by contracting	these cells can change their shape and move to engulf foreign bodies in the bloodstream

2. Some Specialised Plant Cells

epidermal cells	root hair cells	palisade cells
the outer walls have a waxy layer (cuticle) which cuts down water loss	each cell is extended to increase the surface area for water absorption from the soil	have many chloroplasts and are shaped for maximum light absorption

3. Single-Celled Organisms

Some organisms exist as single cells (**unicells**). In these organisms the functions normally carried out by organs and organ systems have to be carried out by structures within the single cell. These structures are called **organelles**. Unicells have a very complex internal structure, compared with bacteria and viruses.

Euglena	*Paramoecium*	*Amoeba*
flagellum (for swimming) chloroplasts (for photosynthesis)	cilia (for swimming) contractile vacuole (to expel excess water)	nucleus food vacuole bacterium being engulfed

4. Animal Cell Ultrastructure

Cells are structured in a highly organised way. Some internal cell structures are called **organelles**.

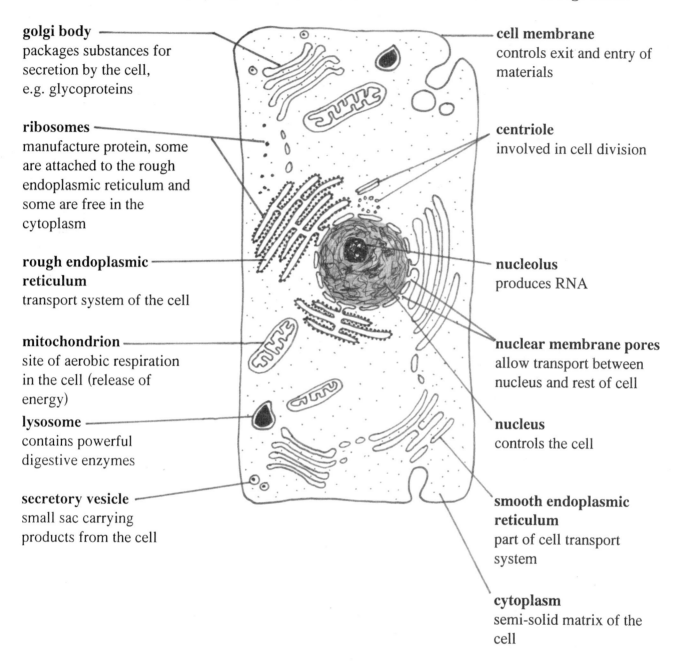

golgi body
packages substances for
secretion by the cell,
e.g. glycoproteins

ribosomes
manufacture protein, some
are attached to the rough
endoplasmic reticulum and
some are free in the
cytoplasm

**rough endoplasmic
reticulum**
transport system of the cell

mitochondrion
site of aerobic respiration
in the cell (release of
energy)

lysosome
contains powerful
digestive enzymes

secretory vesicle
small sac carrying
products from the cell

cell membrane
controls exit and entry of
materials

centriole
involved in cell division

nucleolus
produces RNA

nuclear membrane pores
allow transport between
nucleus and rest of cell

nucleus
controls the cell

**smooth endoplasmic
reticulum**
part of cell transport
system

cytoplasm
semi-solid matrix of the
cell

5. Plant Cell Ultrastructure

All the organelles found in animal cells are also found in plant cells except lysosomes.
Plant cells have in addition:

cell wall	made of slightly elastic cellulose fibres, is fully permeable to water based solutions, provides support and gives shape to the cell
chloroplasts	site of photosynthesis, and contain chlorophyll
large central vacuole	contains a solution of salts and sugars (cell sap) and is a reservoir of water

Absorption and Secretion of Materials

Materials get into and out of cells by five main processes:

1. diffusion   **2.** osmosis   **3.** active transport   **4.** phagocytosis   **5.** pinocytosis

Each of these processes depends on the special properties of the cell membrane.

Structure of the Cell Membrane

The cell membrane is **selectively permeable** and is involved in the control of materials entering and leaving the cell.

The cell membrane is made of **protein** and **phospholipid (fat)** arranged as shown in the diagram below.

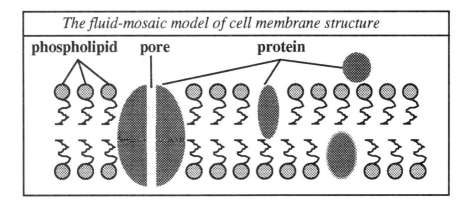

The fluid-mosaic model of cell membrane structure

This structure gives the membrane certain properties:
1. it is easily penetrated by substances which dissolve in fat
2. the proteins give it strength
3. it is a very flexible structure
4. the pores allow chemicals which are insoluble in fat to enter or leave the cell.
5. it is fluid (constantly moving).

Functions of the Cell Membrane

The cell membrane is involved in each of the five main processes by which substances enter and leave cells.

1. Diffusion

Diffusion is the movement of molecules from an area of high concentration to an area of low concentration of those molecules until they are evenly spread. Diffusion is important because it does not use energy supplied by the cell and allows cells to:

- get rid of poisonous waste such as carbon dioxide and urea
- take in oxygen for respiration
- take in food for respiration.

2. Osmosis

Osmosis is the movement of water across a selectively permeable membrane from an area of high water concentration to an area of low water concentration.

The following terms are all associated with osmosis and you should be familiar with their meaning : turgid, flaccid, plasmolysis, hypertonic, hypotonic, isotonic.

3. Active Transport

Diffusion is purely a physical process with molecules moving from a high concentration to a low concentration due to the existence of a **concentration gradient**.
However, the reverse can occur, where molecules move from a low concentration to a high concentration. This movement against a concentration gradient is called **active transport**.

Active transport requires energy produced by respiration. It uses up ATP.
Factors, such as the availability of **oxygen** and the **temperature,** which have an effect on respiration, also have an effect on active transport. Cells which are very active, e.g. liver cells, have a large number of **mitochondria**.

4. Phagocytosis

Large particles (visible under the light microscope) may be taken into a cell by a process called **phagocytosis,** e.g. white blood cells engulfing bacteria.

1. particles engulfed to form vesicle
2. lysosome moves towards vesicle
3. contents of lysosome discharged into vesicle
4. enzymes digest particles and products are absorbed into cell
5. digestion is complete
6. waste products discharged from cell

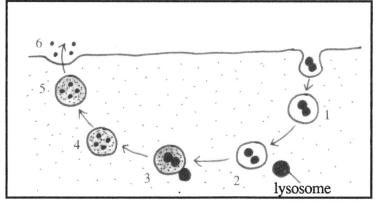

5. Pinocytosis

Pinocytosis is similar to phagocytosis but on a smaller scale and is the method by which liquids, e.g. oil droplets, can be brought into the cell.

1. invagination of membrane forms a vesicle
2. vesicles move into cell
3. smaller vesicles pinch off and move deeper into cell.

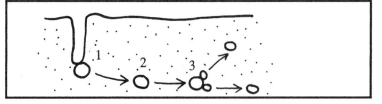

Photosynthesis

The process of photosynthesis involves the manufacture of organic compounds (compounds containing carbon) from carbon dioxide and water, using sunlight as the source of energy and chloropyhll for trapping the light energy.

1. Absorption of Light by a Leaf
Although most of the light striking a leaf is absorbed, only a small part is actually used for photosynthesis.

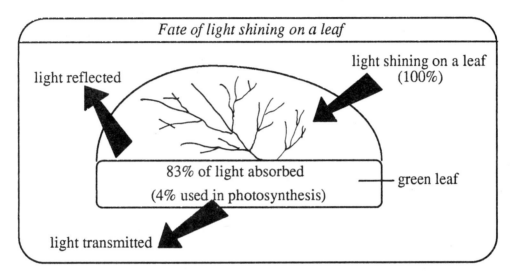

The light that enters a leaf is not all absorbed by chlorophyll. When the absorption spectrum of chlorophyll is examined, only light at the red and blue ends of the spectrum is absorbed. The middle part of the spectrum, green light, is hardly absorbed but reflected, which is why chlorophyll looks green.

2. The Leaf Pigments
Chlorophyll is not just one light-absorbing pigment but four: chlorophyll (a), chlorophyll (b), carotene and xanthophyll.
The pigments absorb light of different wavelengths and pass the energy on to chlorophyll (a) which is the main photosynthetic pigment.
If the percentage absorption by the leaf pigments (**absorption spectrum**) is compared with the photosynthesis rate at different wavelengths (**action spectrum**), there is a close correspondence between the two.

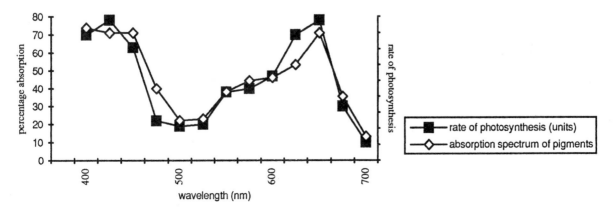

3. The Structure of a Chloroplast
The leaf pigments involved in photosynthesis are contained within the **chloroplasts**.

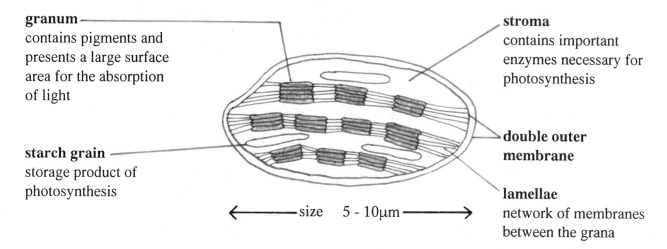

granum
contains pigments and
presents a large surface
area for the absorption
of light

stroma
contains important
enzymes necessary for
photosynthesis

starch grain
storage product of
photosynthesis

**double outer
membrane**

lamellae
network of membranes
between the grana

← size   5 - 10μm →

4. The Chemistry of Photosynthesis
The process of photosynthesis consists of two stages:
- a light-dependent stage involving the photochemical splitting of water (**photolysis**)
- a temperature-dependent stage involving the reduction (combining with hydrogen) of carbon dioxide to form carbohydrate (**carbon fixation**).

5. Photolysis of Water (light-dependent stage)
Occurs in the **grana** of the chloroplast.

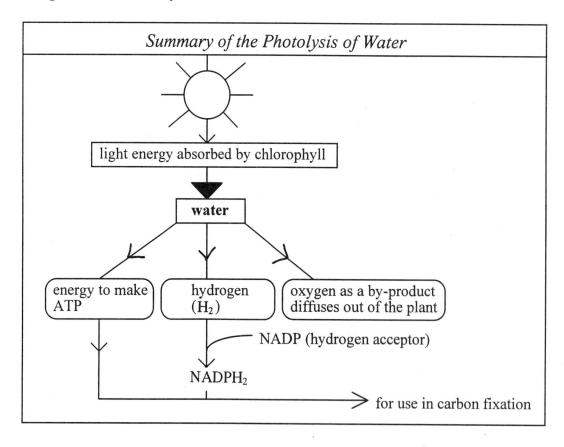

Summary of the Photolysis of Water

light energy absorbed by chlorophyll

water

energy to make ATP | hydrogen (H_2) | oxygen as a by-product diffuses out of the plant

NADP (hydrogen acceptor)

$NADPH_2$

for use in carbon fixation

6. Carbon Fixation (light-independent stage)

Occurs in the **stroma** of the chloroplast.

Carbon fixation involves a sequence of reactions controlled by enzymes, requiring ATP, hydrogen and carbon dioxide. ATP and hydrogen come from the light-dependent stage.

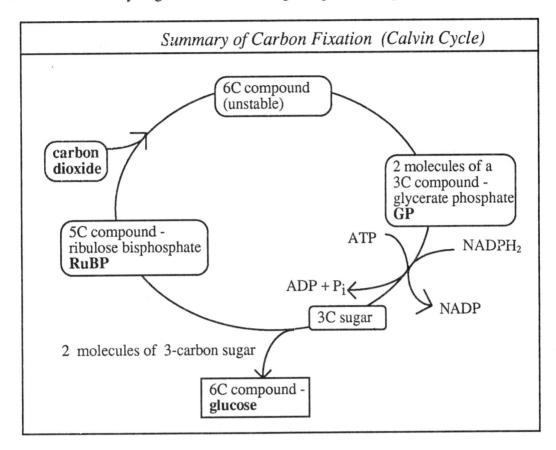

The carbon dioxide enters the chloroplast by diffusion. It combines with a molecule of 5-carbon ribulose bisphosphate to form a 6-carbon molecule. This molecule is unstable and immediately splits into two molecules of 3-carbon glycerate phosphate. Glycerate phosphate is converted to a 3-carbon sugar using hydrogen and energy from ATP. A molecule of glucose is synthesised from two of these 3-carbon sugar molecules. Some of the 3-carbon sugars are needed to regenerate ribulose bisphosphate and are not all used to manufacture glucose.

The photosynthetic process supplies the entire plant with the chemical energy and carbon compounds by which all the major organic molecules, such as carbohydrates, proteins, fats and nucleic acids can be manufactured.

All animals and plants rely directly or indirectly on photosynthesis to supply them with these organic compounds.

Energy Release

Energy is released by the breakdown of organic compounds, such as carbohydrates, fats and proteins. Glucose is the most common respiratory substrate. During respiration, glucose is broken down to release energy which is used to manufacture the chemical **ATP (adenosine triphosphate)**.

1. The manufacture of ATP

ATP is found in all cells and is the universal supplier of energy. Respiration breaks down glucose to release a supply of energy which can be used to manufacture ATP.

The energy which is released during respiration is used to create a bond between the single inorganic phosphate (P_i) and the ADP (adenosine diphosphate).

food and oxygen $\longrightarrow$ ADP + P_i

carbon dioxide + water $\longleftarrow$ ATP

This bond is described as a high energy bond. When it is broken, the energy is released for use by the cell. The energy is therefore stored in the ATP molecule.

Adenosine $— P_i — P_i \sim P_i$

energy rich bond

The energy for the production of ATP comes from two sources:

a)   Some of the steps in the gradual breakdown of glucose are energy-releasing and this energy can be used to produce ATP. Only a small amount of ATP is produced in this way.

b)   The most important source of energy for the production of ATP involves the removal of hydrogen at some of the stages in the breakdown of sugar. The hydrogen is then passed along a system of carrier molecules, finally combining with oxygen to form water. Each time hydrogen is passed from one carrier molecule to another, energy is released and used to produce ATP. The carrier molecules form the **cytochrome system**. NAD is the first carrier molecule.

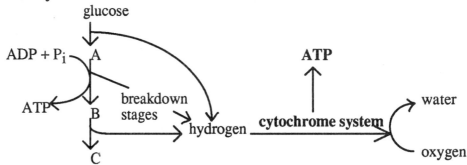

2. The Sites of Respiration

There are three main stages in respiration

- **glycolysis** which occurs in the cell cytoplasm
- **the Krebs Cycle** which occurs in the matrix of the mitochondrion
- **the cytochrome system** (hydrogen carrier system) which occurs on the cristae of the mitochondrion.

mitochondrion structure

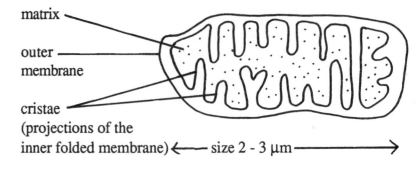

matrix

outer membrane

cristae (projections of the inner folded membrane) $\longleftarrow$ size 2 - 3 µm $\longrightarrow$

The breakdown of glucose is a gradual process involving a number of different stages. Each stage is enzyme-controlled.

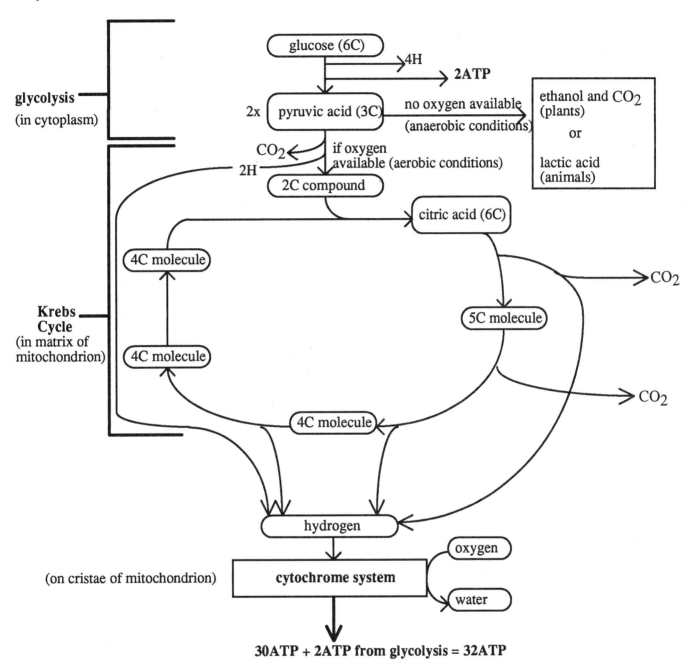

Efficient cells can make use of the 4Hs from glycolysis to build 6ATPs.
The total gain from the entire process in aerobic respiration is **38 molecules of ATP**.

3. Respiration without Oxygen
The majority of energy released during respiration comes from the cytochrome system. This system operates aerobically i.e. oxygen must be available to accept the hydrogen from the final carrier.

Without oxygen (**anaerobic respiration**) the process cannot proceed further than glycolysis with the net gain of only 2 molecules of ATP.

Synthesis and Release of Proteins

Proteins are essential components of all organisms. They play a major role in the regulation of chemical reactions and also form the structural parts of organisms. There are two main types of protein:

- **fibrous proteins**   -   form the main structures of cells and organisms, e.g. the protein keratin is found in hair; collagen is also a fibrous protein
- **globular proteins**   -   enzymes are globular proteins which are essential in the chemistry of organisms; some hormones (chemical messengers) are globular proteins, e.g. insulin, which is involved in the regulation of blood sugar levels.

Everything that a cell does is controlled by enzymes. The nucleus is the control centre of the cell and tells the cell what enzymes to make. It is the chromosomes in the nucleus that contain this information or "blueprint" for an organism.

1. Structure of Chromosomes

Chromosomes are thread-like structures found in the nucleus of a cell. They are made of a chemical called **deoxyribonucleic acid (DNA)**. DNA is made up of basic units called **nucleotides** strung together to form a long chain.

phosphate group

base

5 carbon sugar

a nucleotide

bonds between the sugar and phosphate join the nucleotides together into a strand

There are four different bases: **adenine (A)**, **thymine (T)**, **guanine (G)** and **cytosine (C)**.
To form a DNA molecule, two strands of nucleotides join together with weak bonding forming between their bases. However, only certain bases can join together. Adenine always bonds with thymine and guanine always bonds with cytosine.

* weak hydrogen bonding

A > * > T

C > * > G

T < * < A

the two parallel chains are twisted to form a **double helix**

2. The Replication of DNA

DNA is unique because it is able to reproduce an exact copy of itself. The process is known as **DNA replication.** It involves the following sequence of events:

a) the weak hydrogen bonding between the bases breaks
b) the DNA "unzips" due to enzyme action
c) free nucleotides in the nucleus link up with the appropriate unbonded bases
d) weak bonds form between the bases
e) bonds form between sugar and phosphate groups in adjacent nucleotides
f) 2 new identical DNA molecules are formed.

3. The DNA Code for Proteins

Genes are sequences of bases in DNA. It is genes that carry the instructions for which proteins are to be made by the cell. One gene (a length of DNA) carries the information required to manufacture one protein.

A protein consists of a long chain of amino acids. There are at least twenty different amino acids which can be arranged in any order or number in a particular protein. A typical protein can contain at least 500 amino acids.

It is the sequence of bases on a DNA strand that specifies the proteins to be made by a cell.

A sequence of **3 DNA bases** (**a triplet**) codes for one amino acid.

4. The Synthesis of Protein

The information for the manufacture of a protein has to be carried from the nucleus to the cytoplasm. This is carried out by another nucleic acid called **ribonucleic acid** (**RNA**).

RNA differs from DNA in four ways:

a) it is found in both the nucleus and the cytoplasm
b) it is a single strand, not a double strand
c) it contains the sugar **ribose** (not deoxyribose)
d) it contains the base **uracil** instead of thymine.

There are two types of RNA:

• **messenger RNA** (mRNA)
• **transfer RNA** (tRNA).

5. Protein Synthesis

The part of the DNA molecule to be copied "unzips" and a molecule of mRNA is synthesised alongside it. This copying is called **transcription.**

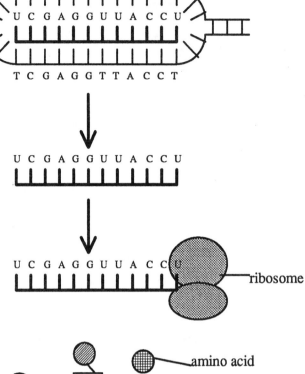

The mRNA leaves the nucleus and passes into the cytoplasm.
The sequence of bases on the mRNA is known as the **codon.**

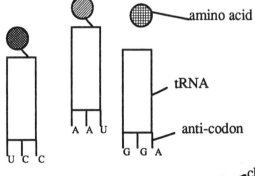

The mRNA becomes attached to a ribosome.

Meanwhile in the cytoplasm tRNA molecules bear a triplet of bases, the **anti-codon.**
This corresponds to a particular amino acid which is picked up by the tRNA.

The ribosome activates the mRNA. Each tRNA anti-codon links with the complementary codon on the mRNA molecule. Bonds form between adjacent amino acids and the protein is built up step by step. The formation of proteins from the DNA code is called **translation.**

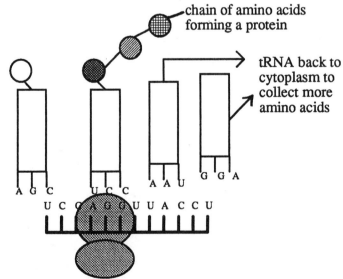

The proteins may be used by the cell or transported to the Golgi Body for packaging and secretion from the cell.

Cellular Response in Defence

Organisms are constantly being invaded by foreign particles, especially bacteria and viruses. Those that cause disease are known as **pathogens**.

1. Invasion of Cells by Viruses

Viruses are too small to be seen through a light microscope - they can only be seen using an electron microscope. The basic structure of a virus consists of a protein coat surrounding a strand of DNA or RNA.

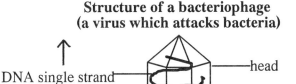

**Structure of a bacteriophage
(a virus which attacks bacteria)**

DNA single strand — head

protein coat

size 0.2μm

tail

Sequence of events when this virus attacks a bacterium:

1. virus approaches cell and sticks to outer membrane
2. the virus tail pierces the cell membrane
3. the viral DNA enters the cell
4. the viral DNA replicates inside the cell
5. new viruses are made from the viral DNA code, the cell bursts (lysis occurs) and the new viruses are released to continue the attack on other bacteria.

2. Cellular Response to Invasion

When an organism is invaded by a pathogen, a variety of responses is initiated.

a) Phagocytosis

Phagocytosis is carried out by some types of white blood cell, e.g. **monocytes**. A monocyte is able to surround a bacterium and ingest it.

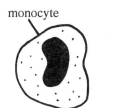
monocyte

bacterium

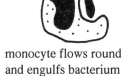
monocyte flows round and engulfs bacterium

lysosomes

bacterium becomes enclosed in a vacuole and enzymes from lysosomes digest it

b) Antibody Formation

Each kind of foreign particle, e.g. polio virus and diptheria bacterium, has chemicals on its surface which are recognised as being "foreign". These chemicals are called **antigens**.

If a particular antigen gets into the body, another type of white blood cell, called a **lymphocyte**, produces a corresponding protein called an **antibody**. These molecules bind onto the antigens. This binding together has two possible effects:

1. It causes the foreign bodies to clump together - this is called agglutination.
2. It may cause them to disintegrate.

In each case these effects are followed by phagocytosis carried out by monocytes.

The production of antibodies in response to antigens is known as the **immune response**.

3. The Immune Response in Transfusions and Transplants
The antigen-antibody reaction is important in blood transfusions and transplants.

a) **Transfusions:** Different blood groups have different antigens on the surface of the red blood cells. The antibodies in the blood plasma are always the opposite of the antigen on the red blood cells. If a person gets blood from the wrong group, e.g. blood group B given blood from group A, the antibodies in the recipient's plasma will agglutinate the red cells of the donor.

blood group	A	B	AB	O
antigen on cells	A	B	AB	none
antibody in plasma	anti B	anti A	none	anti A & anti B

People of blood group AB are known as **universal recipients** because they have no anti A or anti B antibodies in their plasma and can receive blood from anyone without agglutination taking place. People of blood group O are known as **universal donors** because they have no antigens on their red blood cells and can give blood to anyone without causing agglutination.
Other blood grouping systems are also taken into account in transfusions.

b) **Transplants:** If transplants are to be successful then the donor's tissue must match as closely as possible the recipient's, otherwise the transplant will be attacked as a "foreign tissue". Surgeons try to match the antigens of the donor and recipient as closely as possible. This is known as tissue matching/typing. Rejection can be reduced by immuno-suppressant drugs which stop the white cells from attacking the transplant tissue.

4. Plant Responses to Invasion
Attack is mainly through a wound and the response usually occurs at or around the wound. A variety of responses exist, e.g.
* fungal infection can lead to production of large amounts of **phytoalexins** (antifungal chemical)
* the enzyme **chitinase**, which attacks the chitin of insect exoskeletons and the walls of fungal threads, can be produced
* the production of **tannins**, e.g. by potatoes and apples, which can be effective against the protein coat of a virus
* some plants produce **galls** which are abnormal growth patterns which isolate an invading parasite or fungus
* some plants contain large quantities of toxic chemicals such as **nicotine** (e.g. tobacco plant) or **cyanide** (see p48)
* in response to injury many trees and shrubs produce **resin** which in addition to sealing the wound is also distasteful to animals.

2. Genetics and Evolution

Variation

Sexual reproduction is an important source of variation. The mixing of genetic material from two different individuals confers new characteristics on the offspring which may be advantageous.

Variation can arise in three ways:
1. meiosis   - gives rise to gametes which are not all the same
2. fertilisation - fusion of different gametes results in zygotes (fertilised egg cells) with different combinations of genes
3. mutation   - sometimes a change will occur in a gene and a new characteristic may be passed on.

1. Meiosis

Gametes have a single set of chromosomes (one from each homologous pair) and are described as **haploid** (n). A cell with a double set of chromosomes is described as **diploid** (2n). Meiosis is a type of division which results in the production of 4 haploid gametes, e.g. sperm from a diploid mother cell in the testes.

Meiosis involves two consecutive divisions. The diploid mother cell divides into two and these two divide again.

1st Meiotic Division

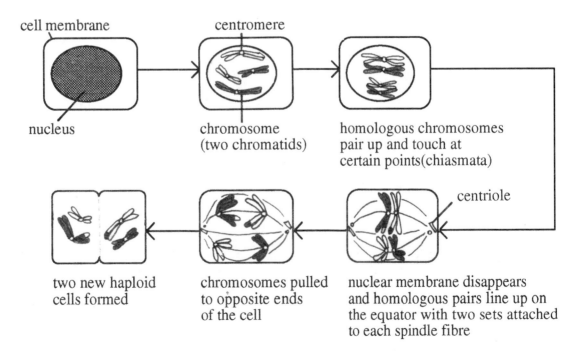

Each of the haploid cells now undergoes a second division resulting in four haploid gametes.

2nd Meiotic Division:

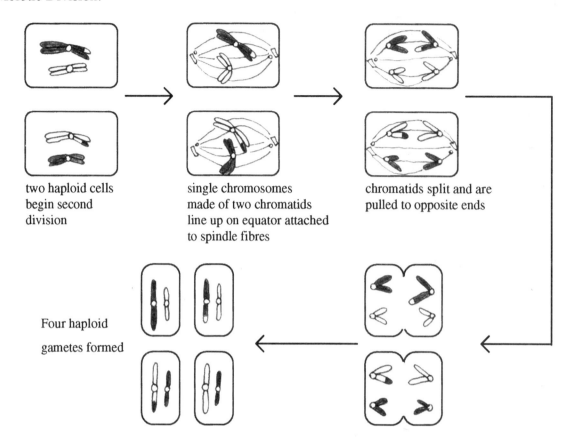

two haploid cells
begin second
division

single chromosomes
made of two chromatids
line up on equator attached
to spindle fibres

chromatids split and are
pulled to opposite ends

Four haploid
gametes formed

Meiosis is a major source of genetic variation. One of the sources is the way in which homologous chromosomes arrange themselves on the equator during the first meiotic division.

2. Independent Assortment

During meiosis the homologous chromosomes come together (assort) but they arrange themselves on the spindle randomly to each other. This is shown in the diagram below which uses as an example only two homologous pairs. There are two ways in which the homologous pairs can arrange themselves.

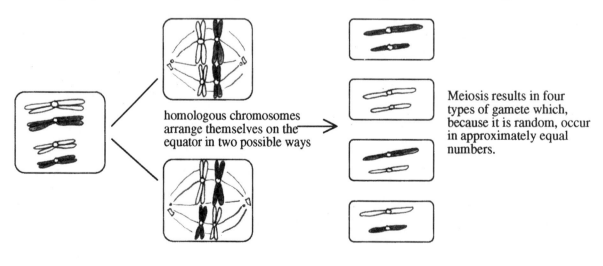

homologous chromosomes
arrange themselves on the
equator in two possible ways

Meiosis results in four
types of gamete which,
because it is random, occur
in approximately equal
numbers.

3. Crossing Over

Meiosis also gives rise to variation through the process of crossing over which occurs at the start of the first meiotic division. Crossing over occurs between adjacent chromatids on homologous chromosomes. Crossing over involves the exchange of genetic material at the points where the chromatids are joined, the **chiasmata**.

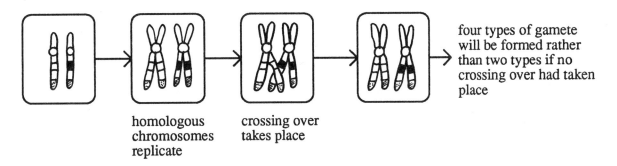

homologous chromosomes replicate

crossing over takes place

four types of gamete will be formed rather than two types if no crossing over had taken place

4. The Dihybrid Cross

The dihybrid cross is a cross between two true-breeding parents which differ in terms of two characteristics. The cross below shows two pea plants which differ in terms of height and flower colour. Let T = tall, t = dwarf, C = coloured, c = white.

parent phenotypes	*tall coloured*	*x*	*dwarf white*
parent genotypes	*TTCC*		*ttcc*
gametes	*TC*		*tc*
F1 genotype		*TtCc*	
F1 self cross	*TtCc*	*x*	*TtCc*
gametes	*TC, Tc, tC, tc*		*TC, Tc, tC, tc*

A **Punnett Square** can be used to show the fusion of gametes.

	TC	*Tc*	*tC*	*tc*
TC	*TTCC*	*TTCc*	*TtCC*	*TtCc*
Tc	*TTCc*	*TTcc*	*TtCc*	*Ttcc*
tC	*TtCC*	*TtCc*	*ttCC*	*ttCc*
tc	*TtCc*	*Ttcc*	*ttCc*	*ttcc*

Expected ratio of phenotypes in F2	*tall coloured*	*tall white*	*dwarf coloured*	*dwarf white*
	9	3	3	1
Possible genotypes	*TTCC, TtCC TtCc, TTCc*	*TTcc, Ttcc*	*ttCC, ttCc*	*ttcc*

5. Linkage

If the inheritance of abdomen width and wing type are studied in *Drosophila*, the normal 9:3:3:1 ratio expected in the F2 generation is not obtained. Instead a 3:1 ratio is obtained. This is due to the genes for these two characteristics being on the same chromosomes. They are said to be **linked** (shown by brackets below).

parent phenotypes	*long wing broad abdomen*	*x*	*vestigial wing narrow abdomen*
parent genotypes	*(LLBB)*		*(llbb)*
gametes	*(LB)*		*(lb)*
F1 genotype		*(LlBb)*	
F1 self cross	*long wing broad abdomen*	*x*	*long wing broad abdomen*
	(LlBb)		*(LlBb)*
gametes	*(LB) or (lb)*		*(LB) or (lb)*

F2 genotypes	*(LLBB)*	*(LlBb)*	*(LlBb)*	*(llbb)*

Ratio of phenotypes in F2:   long wing   :   vestigial wing
                                   broad abdomen   narrow abdomen
                                         3   1

Linked genes do not always stay together. They can be separated as a result of crossing over during meiosis. If a cross-over occurs, it will separate two genes that were previously linked and produce a new combination of genes in the gametes. In this example the genes for colour and shape of kernels are linked i.e. they are found on the same chromosome.

parent phenotypes	*coloured smooth kernels*	*colourless shrunken kernels*
parent genotypes	*(CcSs)*	*(ccss)*
gametes produced if	*(CS) and (cs) along with (Cs) or (cS)*	*all (cs)*
crossing over occurs	*which arose from cross-overs*	

Most of the offspring are of the parental genotypes but a small number are **recombinant** types

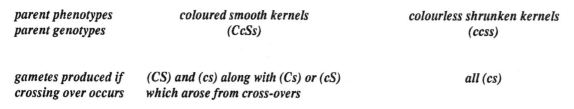

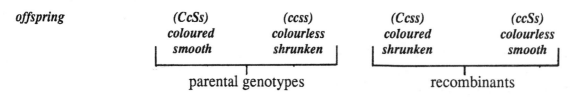

The further apart genes are on chromosomes the more likely crossing over will take place. The **cross over values** are used in producing **chromosome maps**. The greater the number of recombinants the further apart linked genes are on a chromosome.

6. Sex Linkage

In a human female there are two X chromosomes and in a male there are X and Y chromosomes.
All genes carried on the sex chromosomes are said to be **sex linked**.
An example of a sex linked characteristic is **colour blindness**. The allele for colour blindness is recessive and is carried on the X chromosome. Let X^C = normal and X^c = colour blind.

parent phentoypes	*normal woman*	x	*colour blind man*
parent genotypes	$X^C X^C$		$X^c Y$
gametes	X^C		X^c or Y
offspring	$X^C X^c$ *carrier daughter*		$X^C Y$ *normal son*

If one of the daughters has children by a normal man:

	$X^C X^c$	x		$X^C Y$
gametes	X^C or X^c			X^C or Y
offspring	$X^C X^C$ *normal daughter*	$X^C X^c$ *carrier daughter*	$X^C Y$ *normal son*	$X^c Y$ *colour blind son*

The probability of one of their sons being colour blind is 50%.
The probability of one of their daughters being a carrier is 50%.

Haemophilia, a disease in which the blood does not clot properly, is inherited in exactly the same way.
It is also caused by a recessive allele which can be carried on the X chromosome.
Let X^H = normal and X^h = haemophiliac.

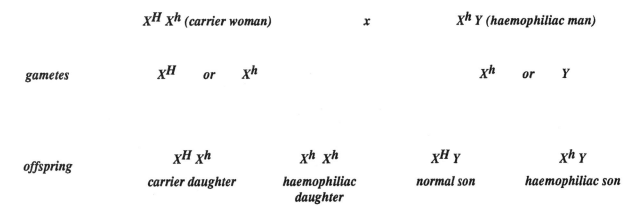

	$X^H X^h$ *(carrier woman)*	x		$X^h Y$ *(haemophiliac man)*
gametes	X^H or X^h			X^h or Y
offspring	$X^H X^h$ *carrier daughter*	$X^h X^h$ *haemophiliac daughter*	$X^H Y$ *normal son*	$X^h Y$ *haemophiliac son*

7. Mutation

Every now and again a population of animals or plants produces an individual which is different from the rest. When this change is due to a change in the genotype, it is called a mutation.

Mutations occur spontaneously and are comparatively rare. Because they cause a change in the genotype of an organism and consequently in the proteins that are manufactured, they are normally harmful or lethal. Mutations can be induced by high temperatures, atomic radiation, ultra-violet light, X-rays or chemicals such as mustard gas.

However mutations are a source of variation, and although most are harmful, occasionally some are advantageous and can contribute to evolution through natural selection as the outcome of a mutation is that a new allele arises in a population.

8. Chromosome Mutations

Involve either a change in structure of a chromosome or a change in the number of chromosomes.

a) Changes in chromosome structure:

1. deletion   - a break occurs and a gene or sequence of genes is lost
2. inversion   - a break occurs and a sequence of genes is rotated 180°
3. duplication   - a sequence of genes is duplicated on a chromosome
4. translocation   - a sequence of genes is moved from one chromosome to another.

b) Changes in chromosome number:

This kind of chromosome mutation involves the addition or loss of one or more chromosomes.

Complete **non-disjunction** can occur during meiosis and is due to homologous chromosomes failing to segregate during the first division.

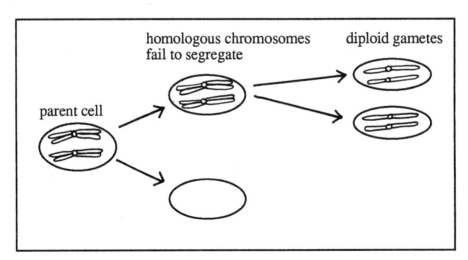

Non-disjunction may affect only one pair of chromosomes, producing sex cells with one chromosome more or less. If abnormal gametes fuse with normal gametes, various conditions arise, e.g. Down's Syndrome caused by non-disjunction of chromosome type 21. If an abnormal egg (n = 24) fuses with a normal sperm (n = 23) the zygote has an extra chromosome type 21 (2n = 47).

The gametes formed by complete non-disjunction are diploid. If two diploid gametes fuse then a tetraploid organism is formed. An organism with extra sets of chromosomes is known as a **polyploid** and the condition is known as polyploidy.

Polyploidy is common in plants which often show increased yield, greater hardiness and increased resistance to disease. An example of polyploidy is shown below.

parents   *Spartina maritima* (2n=56)   X   *Spartina alterniflora* (2n=70)

gametes   28 chromosomes   X   35 chromosomes

$\downarrow$

fusion of haploid gametes gives a
sterile hybrid with 63 chromosomes
(2n=63)
(the hybrid is sterile because the chromosomes
cannot pair at meiosis)

$\downarrow$

failure of separation of chromsomes at mitosis in
the zygote of this hybrid produces a fertile hybrid
(2n=126)
(the hybrid is fertile because the chromosomes
can now pair at meiosis)

This is now a species in its own right called
Spartina townsendii

The chemical **colchicine** can be used to induce polyploidy artificially by preventing the formation of the spindle during mitosis. Since polyploid plants tend to grow larger and more vigorously, many crop plants are improved by inducing polyploidy, e.g. bananas, tomatoes and wheat.

Polypoloidy can also arise when abnormal diploid gametes fuse with one another or with normal haploid gametes.

9. Gene Mutations

A gene mutation involves an alteration in the sequence of nucleotides which will alter the sequence of amino acids coded. This will alter or prevent the formation of proteins and may be lethal. There are four types:

a) substitution-   one or more nucleotides substituted for the correct one
b) insertion   -   one or more nucleotides inserted into the DNA strand
c) deletion   -   one or more nucleotides are deleted
d) inversion   -   nucleotides switched round.

Selection

1. Natural Selection

In an attempt to explain the wide range of animal and plant species, Charles Darwin first published his book "The Origin of Species" in 1859. His theory was based on the evolution of new species being due to a process of natural selection over a long period of time. It involves a number of stages:

a)   A change may occur in the environment either due to an **abiotic** factor such as pollution or a **biotic** factor such as disease.

b)   **Variation** in the population results in some individuals being better adapted to the changing environment.

c)   Competition or 'struggle for existence' favours those individuals adapted to the changing environment - **selective advantage**.

d)   This results in **survival of the fittest** who hand on their favourable characteristics to their offspring.

2. Natural Selection in Action

a)   **The Peppered Moth**

   The peppered moth exists in two forms, a light coloured form and a dark coloured (**melanic**) form. Up until 1880 the dark form was rare. After this time the numbers of the dark form increased dramatically. Before the Industrial Revolution most tree trunks were light in appearance. Towards the end of the 19th century, soot and smoke from the new factories blackened the tree trunks. As a result, in polluted areas the black moths were camouflaged but the light moths were not and were eaten by birds. The dark moths have a selective advantage in polluted areas but in unpolluted areas light moths have the selective advantage. As a result of the Clean Air Acts, the light form is now becoming more common in industrial areas.

b)   **Antibiotic resistance in bacteria**

   When an antibiotic is used, some resistant bacteria will survive and reproduce. This resistance is passed on to their offspring and soon there is a large population of bacteria resistant to the antibiotic. New antibiotics need to be developed as a consequence.

c)   **Resistance to insecticides**

   Due to natural variation some insects will be resistant to insecticides, such as DDT, and will pass on this resistance to their offspring. Once again, a population builds up which is resistant to the insecticide, and other chemicals have to be developed for use as insecticides.

3. Artificial Selection

Since people first started growing crops and domesticating animals, they have selected types which suit their needs. This type of selection is known as **artificial selection**.

Artificial selection involves a careful choice of individuals to be bred. It can involve two types of breeding:

- **inbreeding**   -   closely related individuals are bred, e.g. pedigree dogs from the same family, and this often results in reduced vigour and poor survival rates

- **outbreeding**   -   unrelated individuals are bred resulting in hybrids that are tougher and more fertile - **hybrid vigour**.

Other examples of artificial selection:
- cattle - Jersey, Ayrshire and Guernsey types are bred for milk production and Aberdeen Angus for meat.
- crops - wheat, barley and potatoes bred for higher yield, greater resistance to disease and ability to survive extreme weather conditions.

In any process of selection there will be a loss of a particular organism's genes if they do not reproduce. If a wide variety of species is not conserved then their genes, which may be useful in the future, may be lost. Destruction of environments, such as tropical rainforests, means the loss of many undiscovered species. Loss of diversity can also occur as a result of:
- inbreeding of domestic animals and plants
- the decreasing populations of endangered species.

In order to combat this possible problem, "germ" (sex) cell banks have been set up. Also, rare breed farms are becoming more common. Captive breeding may help prevent **extinction** of **endangered species**.

4. Genetic Engineering

Genetic engineering is the process by which genes are transferred artificially from one organism to another to produce new genotypes, and hence new phenotypes.

Scientists have been able to change the genotype of bacteria to make them produce useful chemicals such as insulin. The technique is known as **recombinant DNA technology**. The technique involves the use of enzymes which act as "molecular scissors", chopping up DNA. These enzymes are called **restriction enzymes**, e.g. endonuclease. Another type of enzyme, **DNA ligase**, joins DNA molecules.

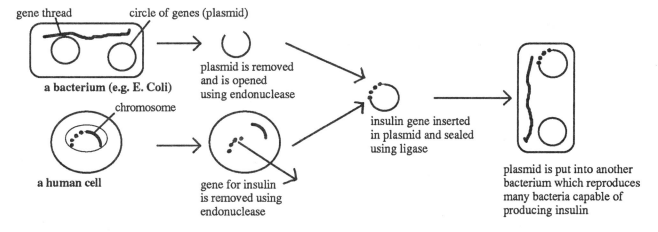

The gene to be removed is identified using gene probes or by recognition of characteristic banding patterns.

Other techniques which are being developed include:
- **somatic fusion of different cells**, e.g. potato and tomato, to produce new varieties of crops with features of both plants
- **somatic fusion of similar plants** with desirable qualities to produce new ones, e.g. wild potato resistant to leaf roll virus fused with domestic variety which is not resistant.

Somatic fusion is used to overcome sexual incompatibility between different plant species. The technique involves the removal of the cell wall by the use of a cellulase enzyme so that the **protoplasts** (plant cells from which the cell wall has been removed) can be fused.

Speciation

1. Formation of a new species
Speciation is the formation of a new species. There are three distinct phases in speciation.

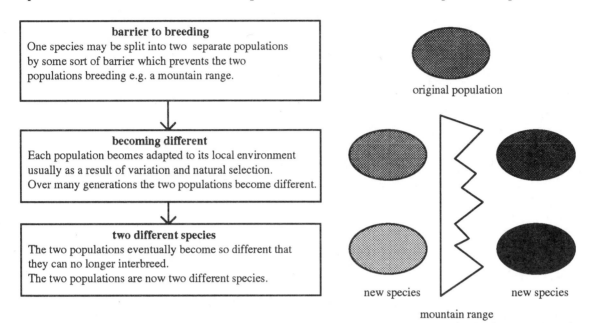

| **barrier to breeding** |
| One species may be split into two separate populations by some sort of barrier which prevents the two populations breeding e.g. a mountain range. |

| **becoming different** |
| Each population beomes adapted to its local environment usually as a result of variation and natural selection. Over many generations the two populations become different. |

| **two different species** |
| The two populations eventually become so different that they can no longer interbreed. The two populations are now two different species. |

original population

new species   new species

mountain range

The barriers which lead to separation of two populations are known as **isolating mechanisms**. These can be:

- **geographical** — oceans, rivers, mountains and deserts
- **ecological** — different habitats may be preferred, e.g. different temperatures, humidities or altitudes
- **reproductive** — caused by two populations being unable to interbreed because of failure of courtship to stimulate, lack of attraction between males and females or failure of gametes to fuse.

2. An example of speciation
On the Galapagos Islands there are 13 species of finch. Each species has a different shape or size of beak. These differences can be explained by natural selection.

| At first there were no finches on the islands until finches from the South American mainland reached them. | The finches displayed variations and adapted to new roles in the new environment. They became different from the original type and from each other. Some flew to other islands. | The variations present in the finches on the other islands allowed them to adapt to their new roles in the environment and they gradually became different species. | Eventually all the islands were colonised, each island having different species of finch. |

From the common ancestral form a radiation of types has occurred. Each island varied in soil, climate and vegetation and eventually finch types adapted to fit the environments. The evolution of all the varieties of types from one common ancestor is called **adaptive radiation**.

3. Control of Growth and Development

Growth Differences between Plants and Animals

1. Growth Patterns

Growth is the permanent increase in size of an organism in the course of its development. There are three processes which contribute to growth:

a) cell division   b) incorporation of raw materials into a cell   c) cell expansion.

The diagrams below show the growth patterns over the lifetime of three different organisms.

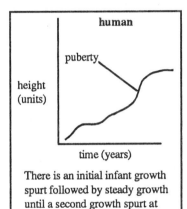

There is an initial infant growth spurt followed by steady growth until a second growth spurt at puberty and then little or no growth after 20.

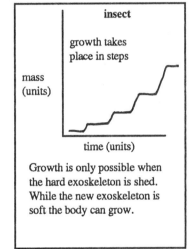

Growth is only possible when the hard exoskeleton is shed. While the new exoskeleton is soft the body can grow.

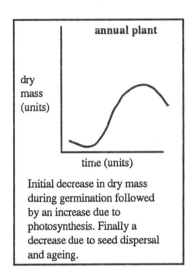

Initial decrease in dry mass during germination followed by an increase due to photosynthesis. Finally a decrease due to seed dispersal and ageing.

2. Growth in Plants

In plants growth and development are achieved by cell division, elongation then specialisation in localised areas called **meristems**. In animals growth takes place all over the body.

The principal meristems are at the tip of the root and shoot and are known as **apical meristems**.

The cells that are produced as a result of mitosis go through a process of growth and **differentiation** to form the permanent tissues of the plant, e.g. xylem, phloem and cortex.

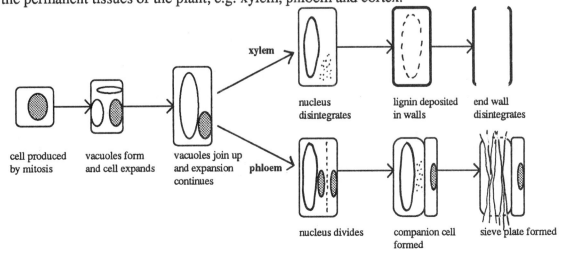

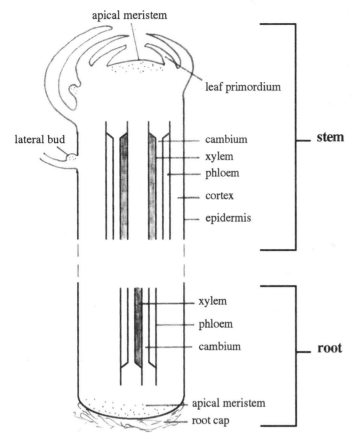

Diagram of a Longitudinal Section through the stem and root of an annual plant

In the shoot the apical meristem is protected by the leaf primordia.

Lateral buds are formed in the angle between the leaves and the main stem.

Lateral buds also have meristematic cells at the tip and are capable of forming side branches.

In the root the apical meristem is protected by the root cap.

3. Secondary Growth in Plants

The activity of the apical meristems increases the length of the shoot and root and forms the permanent tissue of the plant. This is known as **primary growth**. An increase in girth is brought about by a process known as **secondary growth**. Secondary growth takes place in **woody perennial plants** like trees and roses. It does not occur in herbaceous plants.

Secondary growth takes place by division of the meristematic cells located between the xylem and phloem. These cells are called the **cambium**.

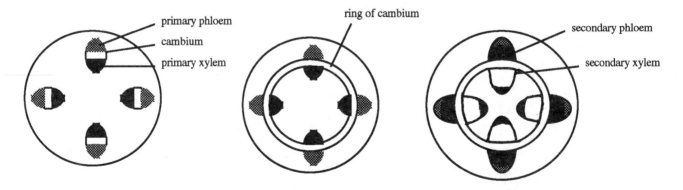

Initially the cambium is restricted to a small group of cells between the xylem and phloem.

The first step involves division of the cambium **radially** to form a ring of cambium round the stem.

The cells of the cambium now divide inwards and outwards to form xylem on the inside and phloem on the outside. Much more xylem is formed so the phloem and cambium get pushed outwards. The xylem and phloem produced is called **secondary xylem and phloem**.

In Britain, the cambium is most active in the spring and inactive in the winter. The spring activity of the cambium produces xylem vessels that are large and have thin walls. This allows for the large flow of water to the leaves during rapid spring growth. The summer wood has smaller, thicker walls.
This seasonal growth of woody plants results in a series of rings being produced. These are the **annual rings** and their size relates to growth conditions, e.g. during a year of drought the ring will be narrower.

In older trees the xylem loses its ability to carry water and becomes filled with gum and resins. This central wood is the **heartwood,** while the outer water-carrying wood is the **sapwood**.

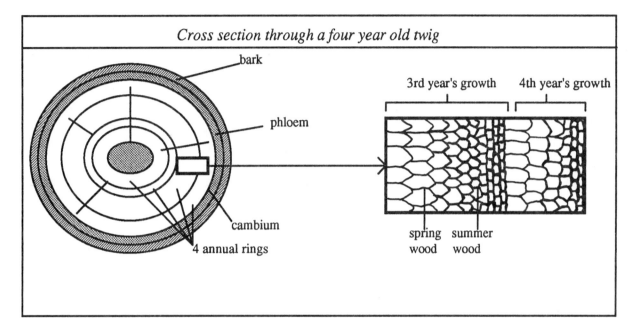

Cross section through a four year old twig

4. Regeneration
Regeneration is the process by which an organism replaces lost or damaged parts.

- Plants: Flowering plants can regenerate whole plants from small fragments. This is known as vegetative reproduction. Small parts can become detached and develop into self - supporting plants, e.g. stem cuttings, pieces of root and even leaves of some plants.
- Animals: The simpler the animals and the less specialised the cells, the greater the ability to regenerate, e.g. a flatworm.
  In complex animals different tissues have different abilities to regenerate. Liver cells have the greatest potential of all to regenerate. Skin cells regenerate all the time, but heart and nerve cells which are highly specialised cannot regenerate.

Genetic Control of Growth

Growth is under the control of both **internal** and **external** factors. External factors include **environmental** factors such as light, oxygen and temperature. Internal factors which affect growth include the **genetic** constitution of the organism and the level of **hormones** in the body.

1. The Role of Genes

Each cell of an organism contains the same genetic information as all others because all cells have developed from repeated divisions of the original single fertilised egg. If all cells have the same genetic information how can we explain:
- muscle cells being able to produce contractile protein and red blood cells being able to produce the oxygen-carrying protein haemoglobin?
- palisade cells being able to produce chlorophyll and cells which will form xylem being able to produce lignin?

Differentiation of cells involves the selective use of the information carried on the genes. Selected genes must be able to work at selected times. External factors such as light or internal factors such as hormones are able to "switch" genes off and on. Differentiation involves interaction between the nucleus and the cytoplasm.

During the late 1950's two scientists, **Jacob** and **Monod**, carried out research investigating this idea of genes "switching" off and on. They carried out a series of experiments involving the bacterium *E.coli* and its synthesis of the enzyme β-galactosidase. This enzyme is involved in the reaction below:

$$\text{lactose} \longrightarrow \text{glucose} + \text{galactose}$$

Their experiments showed that the enzyme β-galactosidase is only produced when it is needed, i.e. when lactose is present. The gene for the production of the enzyme is said to be **induced** by the presence of the specific substrate (lactose) in the environment.

2. The Mechanism for the Lac Operon in *E. Coli*

The structural gene contains the genetic information for the synthesis of the enzyme. It is under the control of the operator gene which is itself under the control of the regulator gene.

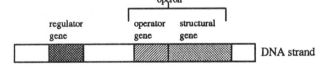

Gene switched off - Lactose absent

When there is no lactose a repressor protein binds to the operator gene and prevents it "switching" on the structural gene.

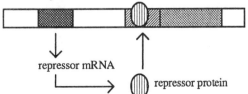

Gene switched on - Lactose present

When the lactose is present it binds to the repressor protein which stops it binding to the operator gene thus the structural gene can function and the enzyme can be produced.

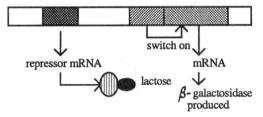

Hormonal Control of Growth

Plants
Plant growth is under the control of a variety of chemical substances called **plant growth substances**.

1. Auxins
The most common auxin is **IAA (indoleacetic acid)**. IAA is produced at the tip of the root and shoot, terminal buds and expanding leaves. It diffuses from the site of production and has a variety of effects on tissues and organs depending on its concentration. Some of the effects are:

- causes cell elongation
- stimulates fruit production
- inhibits growth of lateral buds by auxin diffusing from the terminal bud

- accelerates cell division
- stimulates lateral root formation
- as less IAA reaches the base of a leaf, leaf abscission (break from the main stem) takes place.

IAA is responsible for growth responses to light and gravity.

2. Gibberellins
The most common gibberellin is **GA (gibberellic acid)**. GA also increases cell elongation but, unlike IAA, which has its effect near the tip, GA has its effect at the **internode**. If GA is applied externally to dwarf varieties of plants, it causes internodal growth, resulting in the plant becoming a normal size. Other effects of GA are:

- ends dormancy in buds
- ends dormancy in seeds - GA is released by the embryo of the seed and is transported to the aleurone layer of the seed where it induces the production of the enzyme α-amylase. This breaks the seed dormancy by converting starch to sugar which is used in respiration as the seed germinates.

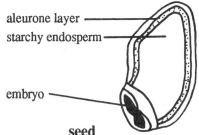

aleurone layer
starchy endosperm
embryo

seed

GA is thought to act at gene level by inducing the production of α-amylase in the cells of the aleurone layer. It binds to the repressor protein and allows the operator gene to switch on the structural gene for α-amylase.

3. Practical Applications of Plant Growth Substances in Agriculture and Horticulture
Plant growth substances can be manufactured synthetically and used to increase production.

Synthetic Auxins
- rooting powder - stimulates production of adventitious roots
- selective weed killer - kills broad leaved plants such as dandelions but grasses are not affected
- pre-harvest drop - synthetic auxins stop the premature drop of fruit and allow easy harvesting
- parthenocarpy - unpollinated flowers treated with auxins can produce fruit
- flowering - can be used to induce flowering in time for seasonal markets, e.g. Christmas.

Synthetic Gibberellins
- malting - treating barley seeds with GA controls the speed of malting (first stage in brewing).

Animals

Animal growth is under the control of chemical substances which act in a similar way to plant growth substances. These chemicals are called **hormones**. Like plant growth substances, they are produced at one site and have their effect at another. Hormones are produced by organs called **endocrine glands** and are released directly into the blood stream.

There are two endocrine glands involved in human growth, the **pituitary gland** and the **thyroid gland**.

1. Pituitary Gland

The pituitary is found at the base of the brain and produces two main hormones involved in growth.

- **growth hormone**   -   promotes growth of the skeleton and muscles

- **thyroid stimulating hormone**   -   has its effect on the thyroid gland and controls the secretions of hormones from the thyroid

2. Thyroid Gland

The thyroid is found in the neck and produces the hormone **thyroxine**.

- **thyroxine**   this hormone controls metabolic rate and has the following effects if production is not normal:

underproduction during development	underproduction in adults	over-production in adults
↓	↓	↓
slow physical and mental development which can be treated by taking thyroxine orally	tissue swelling, decreased metabolic rate and sluggishness which can be treated by taking thyroxine orally	swelling of the thyroid gland and increased metabolic rate which can be treated by surgery

Environmental Control of Growth

1. Minerals and Plants

Plants get all the nutrients they require for normal development from their environment. If an essential nutrient is missing, the plant's development will be affected. As well as needing large amounts of carbon, hydrogen and oxygen to manufacture carbohydrates, plants also need quantities of other chemicals (**macro-elements**) for healthy growth.

mineral required	role in the plant	effect of a deficiency
nitrogen	required for protein manufacture, nucleic acids and chlorophyll	chlorosis (loss of green colour), stunted growth of stem and leaves, reduction in fruit and seed production
phosphorus	important for ATP synthesis and also proteins and nucleic acids	stunted growth, bluish green in colour
potassium	important in cell membrane transport and protein synthesis	stunted growth, margins of leaves are yellowy brown, poor disease resistance
magnesium	needed for the manufacture of chlorophyll	stunted growth and chlorosis

2. Minerals and Animals

Minerals, as well as vitamins, are important for healthy development of animals.

mineral required	role in animal	effect of a deficiency
iron	required for the formation of haemoglobin and is also a component of many enzymes and hydrogen carrying systems	anaemia which is a low red blood cell count
calcium	required for shell, bone and teeth formation, muscle contraction, blood clotting	rickets, osteomalacia and osteoporosis

Some minerals which play no role in growth and development can be present in the environment in high quantities and can have a damaging effect if taken into the body of animals. One example of this is **lead**. Lead can be present in high quantities because:

a) lead water pipes in old buildings can cause lead in drinking water
b) lead enters the atmosphere from the use of leaded fuel in vehicles
c) lead arsenate which is used as a pesticide on tobacco plants can add lead to the atmosphere from tobacco smoke.

Lead is damaging because it inhibits enzymes and has its most damaging effect on young children in whom it can cause brain damage. Lead also affects plant growth as it accumulates in topsoil.

3. Vitamins

Vitamin D has the greatest effect on animal growth. It is essential for the absorption of calcium and phosphorus from the intestine into the blood. These minerals are then used to build bones and teeth. Excess vitamin D is stored in the liver.
There are two sources of vitamin D:

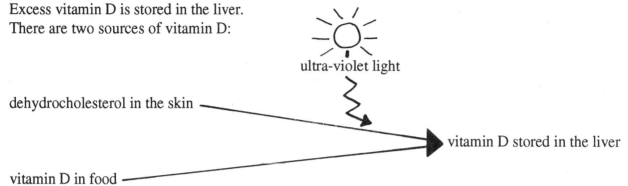

If there is a deficiency of vitamin D in the diet or lack of exposure to sunshine, then deficiency diseases can occur. There are three effects:

a) rickets   -   soft bones cause the legs to bow and there is stunted growth.
b) osteomalacia -   bones become demineralised, especially the pelvis, legs and spine. It is often found in women with poor diets who have had many pregnancies.
c) osteoporosis   -   reduced oestrogen results in the reduction of bone density especially in older women after the menopause, but it can be improved by a healthy diet and regular exercise.

4. Drugs

Many drugs which can have a harmful effect on growth and development can be taken into the body either intentionally, e.g. by smoking, or unintentionally.

- Alcohol   -   As well as having harmful effects on the liver of adults alcohol, if taken by pregnant women, can also have an effect on the growing foetus. It can give rise to a number of birth defects known as **foetal alcohol syndrome**. Some of these effects are:
  - smaller size than normal
  - small heads and brains poorly developed
  - physical defects, e.g. facial abnormalities.

- Thalidomide   -   This drug was prescribed to alleviate morning sickness during pregnancy. It was subsequently discovered that it caused malformed limbs in the growing foetus. The drug had only been tested on non-pregnant adults. It took nearly twenty years to gather scientific proof before victims were compensated.

- Nicotine   -   Evidence now exists showing a connection between smoking and birth defects. Babies born to mothers who smoke during pregnancy may show:
  - high foetal and infant mortality rates
  - abnormalities of the heart and nervous system
  - cleft lip and palate.

5. The Effect of Light on Growth - Plants
a) Shoot Growth
Light has a complex effect on growth. It affects photosynthesis, the production of chlorophyll, the opening and closing of the stomata and growth movements called tropisms. Light suppresses internodal growth and promotes leaf expansion. A plant grown in the dark is yellow in colour, the internodes are long and thin and the leaves reduced. A plant in this condition is termed **etiolated**.

b) Flowering
Many plants respond to a change in the period of exposure to light by producing flower buds. This response to a period of light is called **photoperiodism**. Plants contain a light sensitive chemical which exists in two forms. Each form can change into the other as shown below.

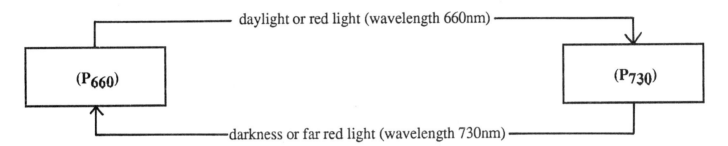

During the day the P_{730} slowly accumulates and during the night P_{660} accumulates.

Flowering is thought to be caused by a hormone. This hormone, it is hypothesised, is produced in the leaves and carried to the flower buds by the phloem.

- Long day plants (e.g. wheat, potato) - These plants only flower when the number of hours of light is above a critical level and the high concentration of P_{730} causes the release of the flowering hormone.

- Short day plants (e.g. primrose, chrysanthemum) - These plants only flower when the number of hours of light is below a critical level and the high concentration of P_{660} causes the release of the flowering hormone. In short day plants flowering can be prevented by interrupting the long dark period with flashes of light which converts some of the P_{660} back to P_{730}. Commercial growers can delay flowering this way.

- Day neutral plants (e.g. celery, tomato) - Photoperiod has no effect on flowering.

6. The Effect of Light on the Timing of Breeding in Animals
Animals also respond to changing periods of daylight (the photoperiod).

- Birds - As daylength increases, the pituitary is stimulated causing the production of sex hormones. The sex organs increase in size and gametes are produced. This prepares the bird for breeding and the rearing of young.

- Mammals - Some mammals such as the hare respond to increasing daylength and breed in the spring with a short gestation period. Other mammals such as the red deer respond to decreasing daylength. They mate in the autumn, with a long gestation period over the winter, and the young are born the next spring.

4. Regulation in Biological Systems

Physiological Homeostasis

All the cells of the body are surrounded by a fluid known as **tissue fluid**. The tissue fluid should not change. The temperature and chemical concentration must be kept constant. This maintenance process is known as **homeostasis**.

1. Control of Blood Sugar Levels

Glucose is one of the main sources of energy and the blood sugar level must be maintained at an optimum level. The rate at which it is released into the bloodstream must be controlled. This means that glucose has to be stored in the **liver,** as glycogen, and released back into the bloodstream under the control of hormones produced by the **pancreas**.

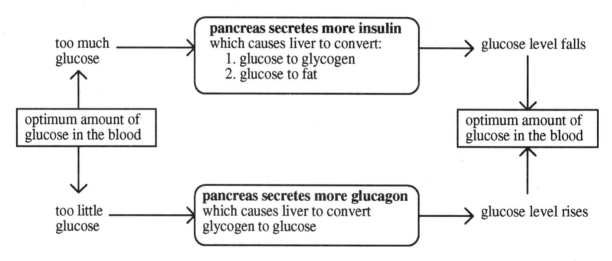

Blood sugar level is increased by the effects of another hormone, which is produced by the adrenal glands (above the kidneys). This hormone is called **adrenaline**.

The control of blood sugar level is an example of a mechanism called **negative feedback**.

In negative feedback a change away from the norm sets a corrective process in motion which brings about a return to the norm. In this case an increase in blood sugar level will bring about a mechanism which causes a decrease.

2. Control of Body Temperature

All metabolic processes are dependent on temperature because of the role of enzymes. Animals which control their temperature internally are known as **endotherms**, e.g. mammals and birds. Animals which cannot control their body temperature internally are known as **ectotherms**, e.g. fish and amphibians.

- Endotherms: The brain plays an important role in temperature control. Changes in body temperature are detected by the **hypothalamus**. A number of effects take place via nerve messages in response to the changing temperature.

As with other homeostatic mechanisms, negative feedback is involved in temperature control.

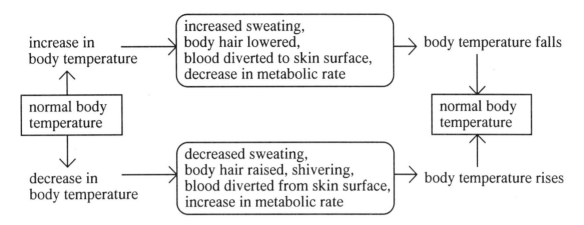

- Ectotherms: Ectotherms often adopt **behavioural methods** to control their body temperature, e.g. lizards will bask in the sun to raise their body temperature, and seek shade, or burrow in the sand, when their body temperature is too high.

3. Control of Water Balance

Blood water level is controlled by the hormone **anti-diuretic hormone (ADH)** which is produced by the pituitary. The hypothalamus of the brain is sensitive to the level of water in the blood. It responds by causing the pituitary to secrete more or less ADH depending on the water concentration.

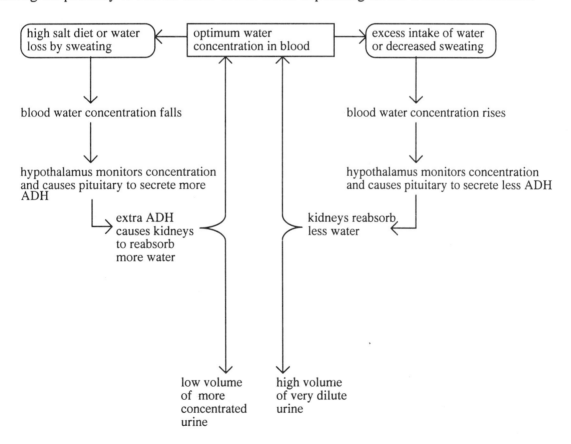

Population Dynamics

1. Population Growth

If a few individuals enter an unoccupied area and there is no shortage of food and no predators, then the population will grow as shown in the graph below.

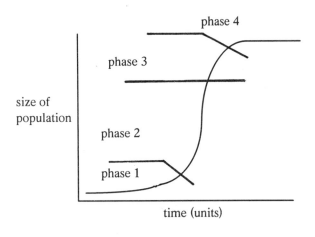

phase 1 - growth starts slowly as there is an initial shortage of reproducing individuals

phase 2 - maximum growth rate under optimum conditions

phase 3 - growth slows down due to density-dependent factors such as shortage of food and space

phase 4 - equilibrium when birth rate = death rate

When a population reaches equilibrium it will not remain constant but will fluctuate because of changes in the surrounding environment. For any population there is an optimum number which can be supported. Any population will tend to fluctuate slightly above or below this optimum.
The fluctuations in a population are caused by many factors. These factors will affect the birth and death rates.

If the population is growing, then the birth rate is greater than the death rate.
If the population is constant, then the birth rate is equal to the death rate.
If the population is falling, then the birth rate is less than the death rate.

The major influences on populations are normally those which cause an increase in the death rate, e.g. food supply, disease, predation, extremes of environmental temperature and breeding space.

2. Factors Influencing Populations

a) Density Independent Factors

Population fluctuations which are not related to variations in the density of a population are called **density independent factors**. These are abiotic factors such as changes in temperature and light. The table below shows that the percentage deaths of three different sized flounder populations were the same during a cold weather spell. The death rate was not dependent on the density of the population.

population	initial population size	number surviving	% deaths
A	16919	1083	93.6
B	55224	2540	95.4
C	2016	135	93.3

b) Density Dependent Factors

Factors which cause fluctuations in the density of a population are known as **density dependent factors** if their action is influenced by the size of the population on which they act. For example, the effects of factors such as disease, availability of food and predation on a population, depend very much on the size of the population.

The table below shows that the percentage survival of Alpine Swift hatchlings varies with the number hatched in each nest.

The greater the number of eggs hatched by the Alpine Swift, the fewer chicks there were that reach fledgling stage.

number of Swift eggs hatched per nest	% reared to fledglings
1	97
2	87
3	79
4	60

3. Predator/Prey Interactions
The diagram below shows how the size of a predator population is affected by the population changes in its prey.

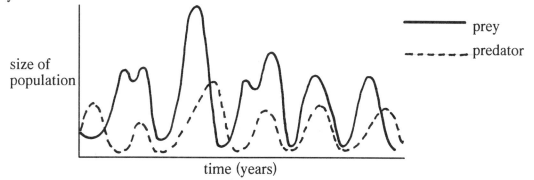

size of population

——— prey

- - - - - predator

time (years)

4. Monitoring Populations
During recent years it has become important to monitor animal and plant populations because of a constantly changing environment. These changes, often due to the activities of people, can have drastic effects on populations which may not be noticed if the populations are not monitored. Examples of these activities include:

- elephants under threat of extinction because of ivory hunters
- reduced fish stocks due to overfishing
- unique habitats threatened, e.g. peat bogs of Sutherland under threat from forestry.

When the level of a pollutant is high, certain organisms may be affected. They may respond with an increase in population size or a decrease. These organisms are known as **indicator species**. Examples are:

- lichens   - sensitive to atmospheric concentrations of sulphur dioxide - fewer lichens in polluted areas
- mayflies   - sensitive to water pollution - numbers fall in polluted water
- birds of prey   - sensitive to pesticide pollution - damages eggs and embryos fail to hatch.

5. Succession and Climax

Communities of animals and plants are not static. Environmental factors may change over a period of time. They may change so much that the environment is no longer suitable for the particular organisms which have colonised it. The colonising organisms themselves may change the environment. Other organisms more suited to the new environment may then move in to take the place of the previous organisms. This gradual change in the plant and animal communities at a particular site is called **succession**. Finally a community develops which does not change. This is called a **climax community**.

Two examples of succession

a) The gradual colonisation of dry rock:

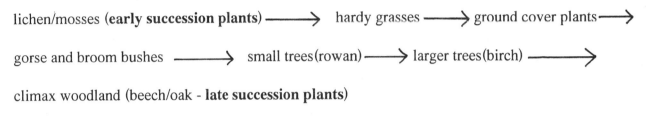

lichen/mosses (**early succession plants**) ⟶ hardy grasses ⟶ ground cover plants ⟶

gorse and broom bushes ⟶ small trees(rowan) ⟶ larger trees(birch) ⟶

climax woodland (beech/oak - **late succession plants**)

b) The colonisation of sand dunes

climax area of ← stable dunes ← mobile sand ← young sand ← area often ← sea
woodland   retaining   dunes -   dunes that   covered by
    water - lichens   marram grass   may be   seawater - sea
    and flowering   covered by   holly
    plants   seawater -
    couch grass

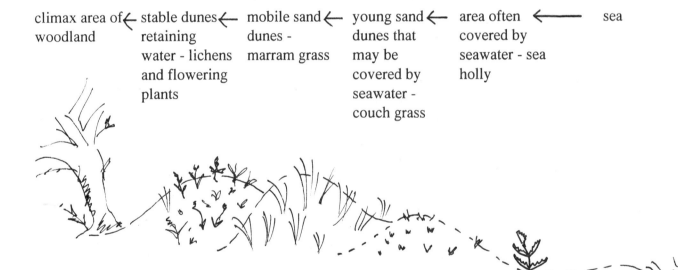

The table below shows the features of early and late succession plants.

Feature	early succession plants	late succession plants
seed germination	helped by light and fluctuating temperatures	not helped by light and fluctuating temperature
photosynthesis rate	high	low
photosynthesis efficiency	low at low light intensities	high at low light intensities
height of vegetation	low	high
species diversity	low	high
feeding relationships	mainly as food chains	mainly as food webs
biomass	low	high

5. Adaptation

Maintaining a Water Balance

Animals

1. Freshwater bony fish, e.g. trout

The body fluids of this type of fish are described as **hypertonic** (more concentrated in terms of salts) to the surrounding water. This means that water tends to enter the cells of tissues, such as the mouth and gills, by osmosis. To counteract this these fish have developed two mechanisms:

a) The kidneys contain many large glomeruli which cause a high filtration rate of the blood leading to the passing of a very large amount of dilute urine.
b) Chloride secretory cells in the gills absorbing salts from the surrounding water to replace the large amounts of salt lost in the large volume of urine.

2. Saltwater bony fish, e.g. herring

The body fluids of this type of fish are described as **hypotonic** (less concentrated in terms of salts) than the surrounding seawater. These fish lose water by osmosis to their surroundings which could lead to dehydration. To counteract this these fish have developed three mechanisms:

a) The kidneys contain very few glomeruli resulting in a very low filtration rate of the blood and little water is lost in the urine.
b) Chloride secretory cells in the gills excreting excess salts, gained by drinking, back into the seawater.
c) To replace lost water, the fish drinks seawater in large volumes.

3. Desert mammals

Animals which live in desert habitats have developed a number of adaptations to ensure survival in these hot, dry surroundings. Some are behavioural and some are physiological. The kangaroo rat is a good example of a desert mammal with these adaptations:

- behavioural   - remains in the cool underground burrow during the heat of the day
- physiological   - does not produce sweat; long kidney nephrons allow maximum reabsorption of water producing a very concentrated urine; dry faeces produced; can survive on water produced by its metabolism.

4. Adaptations associated with salmon and eel migration

Salmon spend the first three years of life in freshwater, migrate to the sea to feed and then return to freshwater to spawn. Eels hatch at sea, migrate to freshwater to complete their growth and then return to sea to spawn. In both cases the fish are moving from surroundings which are hypotonic (freshwater) to surroundings which are hypertonic (seawater) and vice versa. How these fish manage to cope with this is not fully known but it is thought that there must be a change in kidney filtration rate (high in freshwater, low in seawater) and a reversal of the action of the chloride secretory cells (in freshwater, salts are moved in and in seawater salts are moved out). These changes are probably due to genes "switching on and off".

Plants

The flow of water from a plant's roots via the stem to the leaves is called the **transpiration stream**. The plant tissue which allows the transport of water is the **xylem**. The fact that water can be pulled up such long distances relies on the fact that water molecules have a strong attraction to one another (**cohesion**) and to the sides of the xylem vessels (**adhesion**). The water also carries much neeeded minerals to the leaf cells. Transpiration is the evaporation of water from the surface of leaves, mainly through stoma pores. The evaporation of water from the leaves has a cooling effect.

There are a number of factors which affect the rate at which transpiration occurs.

1. Factors Affecting the Rate of Transpiration

factors affecting transpiration rate	*explanation*
temperature	as the temperature rises the rate of water loss increases
humidity	high humidity reduces evaporation as completely saturated air cannot accept any more moisture
wind	wind increases evaporation as the air around the stoma is continually replaced with drier air thus a concentration gradient encourages loss of more water
light	stomata are open in daylight and transpiration rate increases
pollution	air pollution by soot can cause blockage of the stomata thus reducing transpiration

2. Numbers of Stomata

Dicotyledonous plants lose most water through the lower surfaces of their leaves as this is where the majority of stomata are situated. A small proportion of water is lost through the upper surface. Plants which inhabit dry environments have fewer stomata and a reduced rate of transpiration.

3. Stomatal Mechanism

The function of stomata is to allow gas exchange. Stomata open to allow an intake of carbon dioxide and a loss of oxygen. When the stomata are open, water also escapes and the plant may suffer loss of too much water. Controlled opening of the stomata helps solve this problem. Opening and closing of the stomata is controlled by the **guard cells**.

The outer wall of the guard cell is thinner and more elastic. When the guard cells become turgid, the outer walls bulge outwards. The thick inner walls, which are less elastic, are pulled apart. This action causes the pores to open. When the guard cells become flaccid, the reverse occurs and the stomatal pore closes. The pores are open in daylight and closed at night.

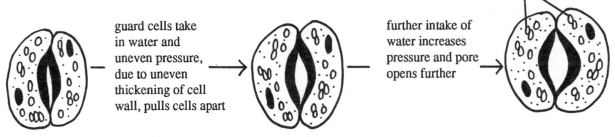

4. Adaptations in Xerophytes

Xerophytes are plants which are adapted for survival in dry environments. Adaptations that these plants may show include:

rolled leaves	*hairy leaves*	*sunken stomata*
Stomata are enclosed on the inside of the rolled-up leaf which reduces water loss as it keeps the conditions within the cavity humid.	Hairs restrict the movement of air over the surface of the leaf thus trapping a layer of humid air close to the leaf surface and reducing transpiration.	Stomata are sunk in deep pits which help to slow down the rate of transpiration.

Other adaptations are:
- small leaves with a small surface area for water loss
- a thick waxy cuticle which reduces water loss
- deep root system to reach water deeper down
- surface roots, covering a large area, which can pick up water when rain falls
- stomata close during the day and open at night
- swollen tissues which can store water.

5. Adaptations in Hydrophytes

Hydrophytes are plants which are adapted for survival in very wet habitats where they may be partly or completely submerged in freshwater. Adaptations which these plants may show include:

long stalks	*floating leaves*	*air spaces in tissues*
The long stalks allow the plant to adapt to changing water levels.	The leaves floating on the surface have stomata on the upper surface.	Air spaces in the stem keep the plant floating near the water surface and near the sunlight.

Obtaining Food

Animals

Animals have to move around in order to find food. This may involve either **hunting** or **foraging** for their food.

1. Foraging Behaviour in Animals

Animals that feed on static food have to move around in order to find it. The type of movement shown by the foraging animals depends on the way in which the food is distributed. The table below describes the behaviour of thrushes when food is evenly distributed or distributed in clumps. The different search patterns are an adaptation which maximises the rate of finding food.

food spacing	*foraging behaviour*	*reason for behaviour*
even	thrushes hop in straight lines after finding food item	when food is evenly spaced the thrush is unlikely to find another food item nearby therefore it is best to keep moving on
in clumps	thrushes make shorter hops and turn sharply after finding food item	when food is clumped it is best to turn in tight angles in order to find the rest of the food items nearby

2. Economics of Hunting Behaviour

Hunting animals must:

- move around in a way that minimises energy expended
- select prey species that maximise the energy gained.

It would be inefficient for animals to hunt for food in a way that expended more energy than the energy value of the food found. Shore crabs choose the mussels which will give them the highest rate of energy return. Very large mussels take so long to open that it is not economic in terms of energy gained. Very small mussels are easy to open but contain very little flesh. There is an optimum size of mussel which gives maximum energy gain for the minimum time spent opening the shell.

3. Competition

There are two types of competition:

a) **Interspecific Competition**   -   Organisms of different species often compete for limited resources. Direct conflict is usually avoided by choosing different prey species or a different feeding area, e.g. different species of wading birds have different lengths of bills and feed at different depths under the mud. They may also feed in different intertidal areas.

b) **Intraspecific Competition**   -   Individuals of the same species often compete for limited resources. This may happen when population levels rise and overcrowding occurs. It can result in aggressive behaviour or migration to other areas.

4. Dominance Hierarchy

Social groups of animals usually involve dominance by some of the animals over others in the group. These dominant individuals have priority over others in access to food, breeding space and mates. Dominance becomes established initially by aggressive behaviour until members of the group recognise each other as those individuals which are stronger and those which are weaker (subordinate), e.g. a pecking order forms in farmyard hens.

Aggression may have become ritualised so that individuals do not fight or come out of a fight relatively uninjured, e.g. stags may crash antlers together but usually little harm is done to either stag.

5. Co-operative Hunting

Animals in social groups may co-operate when hunting prey. This maximises food gain. Groups of hunters can surround prey rather than individuals attempting to chase the prey, e.g. lions hunting antelopes. Large prey species can be subdued more easily by groups.

Once the prey is killed, the dominant animals receive first choice of the dead prey and the subordinate animals must wait for what is left. Co-operative hunting benefits even the subordinate animals since the net energy gain will always be greater than that which could be achieved by lone hunting.

6. Territorial Behaviour

Territorial behaviour is the defence of an area by an individual for the purpose of feeding or mating, e.g. individual humming birds will defend areas of flowers for their exclusive use for feeding. The size of the territory is important: if it is too small it will not provide enough food to ensure survival; if it is too large it will cost too much, in energy terms, to defend it.

Plants

Animals move about in order to find food. Plants are static (sessile) and move very little, if at all.
To obtain food, plants must either:
* make their own food by photosynthesis or
* attract food to them (insectivorous plants attract, trap and digest insects).

Plants compete with each other for light, water and nutrients.

1. Species Diversity

The more species there are in a habitat, the more diverse is the habitat. A diverse habitat is desirable because the more plant species there are:
* the more likely there will be plants of a high nutritional value for grazing animals
* the more animal species can be supported since each will have its preferred food plant
* the more likely it is that there will be plants present which are beneficial to the soil, e.g. nitrogen-fixing plants such as clover.

Low species diversity can result in a reduction of grazing animals and soil nutrients.

Species diversity can be affected by overgrazing. It results in too many species being eaten and a reduction in diversity. Overgrazing of heather moorland results in it reverting to bracken and rough pasture. Undergrazing allows more dominant species to take over, swamping the less competitive. Farmers must continually assess the species diversity of their pasture in order to adjust the level of grazing so that diversity can be maintained.

2. Compensation Point

Plants photosynthesise during daylight. The rate is determined by light intensity. It increases to a maximum at midday when light is brightest, then decreases with decreasing light intensity. Respiration is carried out by plants continually at a more or less constant rate.

The **compensation point** is the point at which the rate of photosynthesis (the manufacture of carbohydrate) and rate of respiration (the using up of carbohydrate) are equal. At this point there is no net loss or gain of carbohydrate.

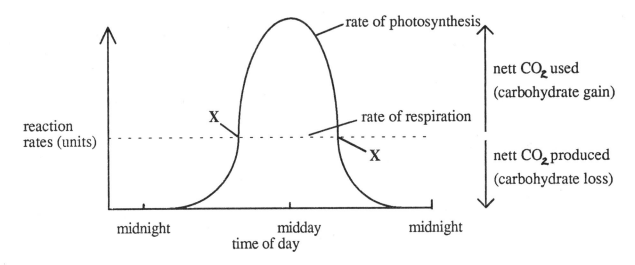

The compensation points are marked **X** on the graph.

3. Sun and Shade Plants

a) **Sun plants**, e.g. deciduous trees such as oak and beech, require high light intensities for photosynthesis. To ensure that maximum light intensity is obtained, sun plants have the following adaptations:

- sturdy erect stems which make the plant tall and able to reach the sunlight
- leaves arranged in a mosaic pattern which avoids overlapping and shading of leaves.

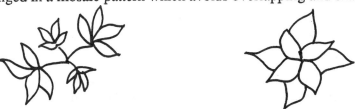

Because they need high light intensities, sun plants tend to have high compensation points.

b) **Shade plants**, e.g. bluebell and primrose, are able to carry out photosynthesis at low light intensities. This ensures their survival in shady environments. These kinds of plant often inhabit woodland floors. Because they can use dim light, shade plants reach their compensation point quickly. They have a low compensation point which allows a net gain of carbohydrate at low light intensity. Such plants often grow early in the season before the leaves of the trees are fully developed.

Coping with Dangers

1. Avoidance Behaviour

Many invertebrates show **escape responses**. These responses bring about a rapid withdrawal from a harmful stimulus. Squid and crayfish dart backwards when faced with danger. The crayfish flexes its abdomen very rapidly and the squid squirts water through its funnel (jet propulsion). Earthworms also show an escape response when they withdraw into their burrows if they are threatened.

2. Learning

Learning is a change in behaviour resulting from past experience. Two forms of learning are:

- **habituation**   If an animal is subjected to a repeated stimulus, it may gradually cease to respond. If the stimulus is not harmful, the animal learns not to react to it and will ignore it e.g. if the ground is tapped in front of a snail, its first response is to withdraw into its shell. Repeated tapping causes habituation and the snail no longer responds to the stimulus. This is an example of short term learning as the normal response will soon return if the stimulus stops.

- **associative learning**   In this type of long term learning an animal learns to associate a particular response with a reward or punishment. Finding food or a mate can be considered a reward and any behaviour which brings about success in feeding or mating will be associated with success and will be repeated.

3. Mechanisms for Defence

individual mechanisms	*social mechanisms*
withdrawal to cover	alarm calls
distraction/diversion displays	grouping of herds/bird flocks
death feigning	fish schooling
chemical defence (warning colouration)	mobbing of intruders

4. Plant Defence Mechanisms

a) Plants which have thorns, spines or stings are less likely to attract grazing animals and are therefore more likely to survive. Because plants are sessile (do not move around) they are vulnerable to grazing.

Within a population of plants, those that have hairier or rougher stems are less likely to be eaten, and will survive and produce offspring. The offspring inherit these characteristics and gradually subsequent generations will consist of higher proportions of individuals showing these characteristics. This is probably how defence structures have evolved in different plant populations.

b) Some plants are able to produce chemical compounds for defence purposes, e.g. wild potato or clover. These chemicals reduce the palatability to grazing animals and they will graze on other plants.

Cyanide compounds, such as those produced by white clover (*Trifolium repens*), are toxic to animals. When leaves are damaged by grazing, cyanogenic compounds are produced rapidly. A non-toxic chemical is produced in one part of the leaf and an enzyme in another. When the leaf is damaged the chemical and the enzyme are mixed producing hydrogen cyanide, a highly toxic poison.

The ability to produce cyanogenic compounds is determined genetically - some plants have the ability, some do not. Where the two types exist together, grazing animals will select the acyanogenic plants (plants without the ability to produce the toxin). Thus the cyanogenic plants will survive and the numbers in the population increase.

c) Most plants grow from meristems well above the ground. If they are damaged during grazing, regeneration is difficult. The meristems of grasses are at ground level and are unlikely to be damaged even after heavy grazing. New leaves grow quickly and thus grasses are better able to tolerate grazing.

Deep root systems and underground stems also help some plants to tolerate grazing.

Printed by Inglis Allen, Kirkcaldy